YOUR KNOWLEDGE HAS VALUE

- We will publish your bachelor's and master's thesis, essays and papers

- Your own eBook and book - sold worldwide in all relevant shops

- Earn money with each sale

Upload your text at www.GRIN.com and publish for free

Bibliographic information published by the German National Library:

The German National Library lists this publication in the National Bibliography; detailed bibliographic data are available on the Internet at http://dnb.dnb.de .

Imprint:

Copyright © 2017 GRIN Verlag
Print and binding: Books on Demand GmbH, Norderstedt Germany
ISBN: 9783668628090

This book at GRIN:

https://www.grin.com/document/388767

Roland Nathan Kalonji

Improving the Application of the "Learning Curve"

GRIN Verlag

Application of Learning Curve theory

<u>REPORT</u>

By

Roland Nathan Kalonji

<u>19-07-2017</u>

Table of Contents

List of figures

List of Tables

Abstract

There exist some fields in which the learning curves can be applied; they have a much wider applications including manufacturing and marketing strategy. However, they underlay the concept of continuous improvement, pricing decisions, work scheduling, standard setting, direct labour budget etc. Learning curve states that decreasing man hours are required to accomplish any repetitive task as the operation is continued; then knowledge of learning curve can be useful both in planning and control.

Estimates of learning curves are used in many applications in organizations and they can be effective with the job which is repetitive in nature particulars with same machinery and tools; thus this study aims at improving the application of the learning curve which describes how knowledge is acquired and retained when people repeat a process. The overall purpose of this report is to determine whether or not the learning curve theory can be replicated.

Data will be gathered based on the England flyer folding instructions; thereafter we will be plotting the learning curve by calculating the cumulative average, estimating the table of unit improvement curves and determining the learning rate.

1. Introduction

1.1 Background and motivation

When a new product or process is started, performance of worker is not at its best and learning phenomenon takes place. As the experience is gained, the performance of worker improves, time taken per unit reduces; and then his productivity goes up. Thus, people who learn always increase their experience, they gain skill and ability from their own experience and the results throughout processes are improved. The usual use of learning curves is to estimate the labor hours and thereby much of the cost, of a manufactured good that is built in significant quantities.It can be applied to the individuals or organisations based on several assumptions:

- Standardised product: the product isn't changing; then every time the product changes, the learning effect will stop and start again.
- Little or no breaks in production & little or no labour turnover.
- The unit time will decrease at a decreasing rate.
- Complex operation: simple operation doesn't require learning.
- The reduction in time will follow a predictable pattern.
- The process is labour intensive: the labour determines the speed of the process.

The learning curve is a line displaying the relationship between *unit production time* and *the cumulative number of units produced*. Learning curves have been applied to creation of documents, boring of tunnels, drilling of wells, upgrades of previously manufactured products, and many other repetitive activities This time around, two learning curve models are in widespread use: the "unit" (U) model, due to Crawford, and the "cumulative average" (CA) model, due to Wright. Some manufacturing companies also apply "learning" to the purchase of raw materials and also to the purchase of manufactured components from other companies (Evin Stump P.E). Then, an organization also acquires knowledge in its technology, its structure, documents that it retains, and standard operating procedures [1].

Applications of the learning curve:

- **Internal**: determine labor standards and rates of material supply required.
- **External**: determine purchase costs.
- **Strategic**: determine volume-cost changes.

1.2 Problem statement

For purposes of the current study, an assignment has been set to determine whether or not the learning curve theory can be replicated.

2. Literature Review

<u>The method study</u> was originally designed for the *analysis* and *improvement of repetitive manual work*; though it can be used for all types of activity at all levels of an organisation, is the process of subjecting work to systematic, critical scrutiny to make it more effective and or more efficient. The aim of method study is to *analyse a situation, examine the objectives of the situation* and then *to synthesize an improved, more efficient and effective method or system* [2].

In fact, it is one of the keys to achieving productivity improvement whereby the process is often seen as linear. The basic procedure was first developed and articulated by Russell Currie at Imperial Chemical Industries (ICI) and consists of six steps (SREDIM):

- <u>Select</u> (the work to be studied);
- <u>Record</u> (all relevant information about that work);
- <u>Examine</u> (the recorded information);
- <u>Develop</u> (an improved way of doing things);
- <u>Install</u> (the new method as standard practice);
- <u>Maintain</u> (the new standard proactive).

2.1 Learning Curve

The learning curve is based on a *doubling of production*: That is, when production doubles, the decrease in time per unit affects the rate of the learning curve. Mostly, cost is related to time or labor hours consumed, learning curves are very important in industrial cost analysis. A key idea underlying the theory is that every time the production quantity doubles, we can expect a more or less fixed percentage decrease in the effort required to build a single unit (the Crawford theory).

The "learning effect" was first noted in the 1920s in connection with aircraft production. It use was amplified by experience in connection with aircraft production in WW II. Initially, it was thought to be solely due to the learning of the workers as they repeated their tasks. Later, it was observed that other factors probably entered in, such as improved tools and working conditions, and various management initiatives.

Learning curve was first described by psychologist Hermann Ebbinghaus in 1885 and is used as a way to measure production efficiency and to forecast costs. Afterward, the learning curves were applied to industry in a report by T. P. Wright of Curtis-Wright Corp. In 1936 Wright described how direct labor costs of making a particular airplane decreased with learning, a theory since confirmed by other aircraft manufacturers. Regardless of the time needed to produce the first plane, learning curves are found to apply to various categories of air frames [3].

Furthermore, it has since been applied not only to labor but also to a wide variety of other costs, including material and purchased components; and the power of the learning curve is so significant that it plays a big role in many strategic decisions related to employment levels, costs, capacity, and pricing.

The Learning curves are mathematical models used to estimate efficiencies gained when an activity is repeated; the principle is shown as: $T \times L^n$ = Time required for the nth unit
Where T = unit cost or unit time of the first unit
$\quad L$ = learning curve rate
$\quad n$ = number of times T is doubled

For instance, if the first unit of a particular product took *10 labor-hours*, and if a *70%* learning curve is present, the hours the fourth unit will take require doubling twice; from 1 to 2 to 4.
Thus, the formula is: **Hours required for unit 4 = 10 × (.7)2 = 4.9 hours**

Industry learning curves vary widely and have various effects; and different organizations, products, have different learning curves. The rate of learning varies depending on the quality of management and the potential of the process and product. Any change in *process, product, or personnel* disrupts the *learning curve*. Thus, caution should be exercised in assuming that a learning curve is continuing and permanent.

Table 1 Example of Learning-curve effects

	EXAMPLE	IMPROVING PARAMETER	CUMULATIVE PARAMETER	LEARNING CURVE SLOPE (%)
1	Model-T production	Price	Units produced	86
2	Aircraft Assembly	Direct labor-hours per unit	Units produced	80
3	Equipment maintenance at GE	Average time to replace a group of parts	Number of replacements	76
4	Steel production	Production worker labor-hours per unit produced	Units produced	79
5	Integrated circuits	Average price per unit	Units produced	72
6	Hand-held calculator	Average factory selling price	Units produced	74
7	Disk memory drives	Average price per bit	Number of bits	76
8	Heart transplants	1-years death rates	Transplants completed	79

2.2 Learning Curves Approaches

A. Arithmetic Approach

The arithmetic approach looks to be the simplest approach to learning-curve equations; each time that production doubles labor per unit declines by a constant factor, known as the learning rate. Thus, if we the learning rate is 80% and that the first unit produced took 100 hours, the hours required to produce the second, fourth, eighth, and sixteenth units are as follows:

Table 2 Learning curve-Arithmetic Approach

NTH UNIT PRODUCED	HOURS FOR NTH UNIT
1	100
2	$80.0 = (.8 \times 100)$
4	$64.0 = (.8 \times 80)$
8	$51.2 = (.8 \times 64)$
16	$41.0 = (.8 \times 51.2)$

The hours required to produce N units and N is one of the doubled values. But, the Arithmetic approach does not show how many hours will be needed to produce other units. For this flexibility, the logarithmic approach seems useful.

B. Logarithm Approach

The logarithmic approach determines labor for any unit, TN, by the formula:

$TN = T1\ (Nb)$

Where TN = time for the Nth unit

$\quad\quad T1$ = hours to produce the first unit

$\quad\quad b$ = (log of the learning rate)/ (log 2) = slope of the learning curve

For instance, the learning rate for a particular operation is 80%, and the first unit of production took 100 hours. The hours required to produce the third unit may be computed as follows:

$TN = T1\ (Nb)$

$T3 = (100 \text{ hours}) (3^{b})$

$\quad = (100) (3^{\log.8/\log 2})$

$\quad = (100) (3^{-.322}) = 70.2 \text{ labor-hours}$

C. Coefficient Approach

The learning-curve coefficient is shown by the following equation:

$$TN = T1C$$

Where TN = number of labor-hours required to produce the Nth unit

$\quad\quad T1$ = number of labor-hours required to produce the first unit

$\quad\quad C$ = learning-curve coefficient

The learning-curve coefficient, C, depends on both the learning rate and the unit of interest.

Table 3 Learning Curve-Coefficient Approach

UNIT NUMBER (N)TIME	70%		75%		80%		85%		90%	
	UNIT TIME	TOTAL TIME	UNIT TIME	TOTAL TIME	UNIT TIME	TOTAL TIME	UNIT TIME	TOTAL TIME	UNIT TIME	TOTAL TIME
1	1	1	1	1	1	1	1	1	1	1
2	0.7	1.7	0.75	1.75	0.8	1.8	0.85	1.85	0.9	1.9
3	0.568	2.268	0.634	2.384	0.702	2.502	0.773	2.623	0.846	2.746
4	0.49	2.758	0.562	2.946	0.64	3.142	0.723	3.345	0.81	3.556
5	0.437	3.195	0.513	3.459	0.596	3.738	0.686	4.031	0.783	4.339

2.3 The Standard time and Normal time

A. Standard Time

The standard time is the time required by an average skilled operator, working at a normal pace, to perform a specified task using a prescribed method; it's also known as the time that a qualified and well trained operator working at a normal pace will need to complete one cycle of the operation.

The standard time includes appropriate allowances to allow the person to recover from fatigue and, where necessary, an additional allowance to cover contingent elements which may occur but have not been observed.

> **Standard time = normal time +allowance**

The Standard Time is the product of three factors:

- **Observed time**: The time measured to complete the task.
- **Performance rating factor**: The pace the person is working at; 90% is working slower than normal, 110% is working faster than normal, 100% is normal.
- **Personal**, **Fatigue**, and **Delay** (PFD) allowance.

And the standard time can be determined using the following techniques: Time study, predetermined motion time system (PMTS or PTS), Standard data system, work sampling and Method of calculation. It can also be used in several branches such as : Evaluation of alternative methods, Labor cost control and manpower planning, overhead cost estimation and budgeting, production scheduling: CPM, Production line balancing, Plant layout and plant capacity, Training and performance evaluation, Output-based incentive scheme design .

B. Normal Time

The normal time (base time or levelled time), is the time required by a trained worker to perform a task at a normal pace. And, the total of all the normal elemental times constituting a cycle or operation.

> **Normal time = avg time *rating factor**

3. Research Question

Although the study consists on a method through many processes, the question is to determine whether or not the learning curve theory can be replicated.

4. Objectives

The objectives of this study are:

- To get the learning curve through a logarithm analysis.
- To estimate the learning rate and the improvement curves.

5. Experimentation

5.1 Methodology
The following procedure has been followed in order to get the learning curve:

1 selecting the design

2 Setting up the apparatus (stop watch, etc.)

3 Looking at the 'England flyer' folding instructions

4 Breaking down the time (Time assigned for each folding step)

5 Starting the experiment

5.2 Apparatus
The apparatus used are as depicted below:
- Papers
- Pen
- Stopwatch
- Ruler
- Computer
- Calculator

6. Observations

The raw data are as depicted below:

Table 4 Raw data

5.24	4.8	4.06	3.5	4.52	3.03	3.64	5.3	4.7	5.5	5.3	5	9.5
13.57	12.9	11.1	12.8	7.95	10.04	8.64	7.4	9.42	10.9	8.1	8.5	9.4
11.68	11.5	6.79	3.8	6.25	4.75	6.63	4.3	5.38	6.2	5.1	9.2	4.6
18.59	11.9	8.77	9	9.75	7.87	9.27	11.4	6.74	10.5	9.1	9.9	5.8
11.77	12.7	8.19	7.7	7.05	4.93	8.13	8.8	8.26	7	11.7	8.6	5.8
19.69	14.9	6.4	8.6	3.03	3.55	10.15	7.8	4.69	8.1	6.4	8.5	7.8
18.36	14	12.64	10.6	17.6	9.94	8.84	7.5	9.28	5.9	5	4.4	5.9
6.09	5	3.61	6	4.91	4.64	2.12	6.8	2.82	7.3	6.4	4.4	4.6
6.9	5.9	3.74	5.4	2.73	4.72	3.41	5.6	4.02	3.8	5	2.5	4.4
10.19	9.5	7.97	4.8	7.47	7.95	5.82	3.5	6.35	2.6	3.9	3.3	2.2
11.15	5.2	9.05	7.1	6.48	10.75	5.59	4	9.58	4.2	4.7	4	6
133.23	108.3	82.32	79.3	77.74	72.17	72.24	72.4	71.24	72	70.7	68.3	66

4.8	4.9	3.7	3.8	4.3	4.44	3.79	5.27	6.6	4.8	9.6	4.2	3.9
8.4	7.5	9.5	9.4	9.3	7.27	6.69	8.55	15	5.6	6.2	10.5	11.4
5.1	5.9	4.9	5.9	5.3	4.7	7.17	3.19	4.5	7.8	4.4	6.5	5.2
8.3	8	6.5	7.2	6.3	4.84	6.85	5.98	6.5	8.1	6	5.9	5.9
7.2	8	9	7	9.1	6.77	5.28	7.6	4.9	6.1	7.6	6.9	5.8
8.3	6.3	5.8	6.9	5.6	4.56	5.14	4.02	6.1	6.6	7.9	7.5	5.5
6	6.4	5.8	7.1	6.8	9.72	7.82	8.74	6.1	6.2	4.1	4.1	5.9
6.3	6.3	4.3	4.5	5.8	4.21	2.98	2.4	3.4	3.5	5	4.9	5
5.5	4.8	4.7	4.5	4.2	3.37	3.11	6.5	3.2	3.7	3.4	3.5	4.1
2.6	3.2	2.6	2.5	2	5.52	7.27	5.33	2.4	3	2.5	1.5	2.5
4.1	3.1	5.3	3.4	4.1	4.71	3.56	1.55	58.7	55.4	56.7	55.5	55.2
66.6	64.4	62.1	62.2	62.8	60.11	59.66	59.13					

4.9	5.9	4.53	3.99	4.9	7.4	4.4	4.13	3.61	2.98	4.2	7.1	4
12.4	5.2	8.19	7.6	9	7.7	6.4	7.14	7.03	6.53	7.3	4.4	4.5
7.8	5	4.52	4.45	3.7	4.2	5	4.15	4.52	4.34	3.37	5	5.5
5.9	6.7	7.94	7.69	5.4	5.6	5.9	5.6	6.53	5.55	6.5	7.7	9.1
5.6	7	5.9	6.94	6.1	6.1	6.7	4.21	5.44	4.71	5.13	7.4	5.4
4.9	10.3	3.75	4.3	7.5	5.4	5.4	2.9	3.35	3.92	3.44	3.3	7.5
5.3	4.4	6.67	6.81	5.6	5.2	6	9.36	7.92	8.93	7.7	4.4	4.5
3.5	5.3	2.55	2.62	3.8	5.3	6.4	2.24	2.55	2.73	2.81	5.2	5.3
3.2	3.8	2.8	1.58	3.9	3.8	2.9	2.22	2.64	2.04	2.35	3.7	3.4
1.9	2.6	4.31	8.28	2.7	1.9	3.4	5.63	4.13	3.64	4.23	2.8	2.4
55.4	56.2	4.71	54.26	52.6	52.6	52.5	4.27	3.95	6.01	4.6	51	51.6
		55.87					51.85	51.67	51.38	51.63		

4.4	3.9	3.76	3.55	4.03	3.97	4.2	3.9	2.8	3.53	3.64	3.42	4.06
12.1	8.4	9.14	6.15	6.36	9.14	3.6	8.4	7.6	7.62	6.47	7.74	6.27
2.1	3.9	4.76	3.35	3.83	4.05	5.3	3.9	3.7	3.7	3.47	3.53	3.94
5.6	6.4	6.17	5.52	5.64	5.68	6.4	6.4	5.8	5.47	4.89	6.55	4.76
5.8	6.5	3.33	7.71	6.96	3.65	6	6.5	7.3	5.07	5.93	4.94	4.99
6.8	5.4	3.15	5.22	2.69	4.36	8.2	5.4	5.7	3.95	4.02	5.29	4.07
5.2	5.7	6.75	6.52	9.04	6.81	5	5.7	6.2	6.39	8.2	9.35	6.21
5	4.9	2.73	2.93	2.28	3.21	5.6	4.9	4.9	2.54	2.41	2.63	2.34
2.6	3.7	2.24	2.67	1.37	1.65	3.8	3.7	3.8	3.19	1.82	5.24	2.23
2.5	2	3.67	3.35	3.33	3.49	2.4	2	1.5	3.81	3.41		4.32
52.1	50.8	4.41	3.99	5.28	4.29	50.5	50.8	49.3	4.09	5.4	48.69	5.18
		50.11	50.96	50.81	50.3				49.36	49.66		48.37

4.62	4.61	3.5	6	3.8	4.49
5.2	5.8	5.98	7.7	5.7	6.74
3.63	2.86	4.72	4.3	4.6	4.15
6.04	5.03	4.43	1	4.2	5.34
5	5.97	4.76	7.3	5.7	5.86
3.52	3.13	3.78	5.5	5.7	2.61
6.7	7.79	7.26	5.6	5.8	6
2.85	2.15	2.08	4	5	2.33
1.93	2.13	2.1	3	3.7	1.39
3.65	3.9	4.02	2.4	1.8	3.67
4.68	4.3	4.88	46.8	46	42.58
47.82	47.67	47.51			

Table 5 Raw data (Unit & time)

Unit	Time	Unit	Time
1	133.23	31	52.6
2	108.3	32	52.6
3	82.32	33	52.5
4	79.3	34	51.85
5	77.74	35	51.67
6	72.17	36	51.38
7	72.24	37	51.63
8	72.4	38	51
9	71.24	39	51.6
10	72	40	52.1
11	70.7	41	50.8
12	68.3	42	50.11
13	66	43	50.96
14	66.6	44	50.81
15	64.4	45	50.3
16	62.1	46	50.5
17	62.2	47	50.8
18	62.8	48	49.3
19	60.11	49	49.36
20	59.66	50	49.66
21	59.13	51	48.69
22	58.7	52	48.37
23	55.4	53	47.82
24	56.7	54	47.67
25	55.5	55	47.51
26	55.2	56	46.8
27	55.4	57	46
28	56.2	58	42.58
29	55.87	59	42.16
30	54.26	60	41.58

7. Analysis

In view to determine the scale of enhancement for each time and the decreasing part, we need to calculate the cumulative average before the eventual results; the sample calculations for CA in table 6 is as follows:

- CA (row 0 ,unit 1,3rd column) = 133.23 × 1 = 133.23

- CA (row 1,unit 2, 3rd column) = $\dfrac{\mathbf{133.23 + 108.3}}{\mathbf{2}}$ = 120.765

- CA (row 2,unit 3, 3rd column) = $\dfrac{\mathbf{120.765 + 82.32}}{\mathbf{2}}$ = 101.5425

- CA (row 3,unit 4, 3rd column) = $\dfrac{101.5425 + 79.3}{2}$ = 90.42125

- CA (row 4, unit 5, 3rd column) = $\dfrac{90.42125 + 77.74}{2}$ = 84.080625

- CA (row 5, unit 6, 3rd column) = $\dfrac{84.080625 + 72.17}{2}$ = 78.1253125

- CA (row 7, unit 7, 3rd column) = $\dfrac{78.1253125 + 72.24}{2}$ = 75.18265625

- CA (row 9, unit 8 ,3rd column) = $\dfrac{75.18265625 + 72.4}{2}$ = 73.79132813

- Etc.

Table 6 Cumulative averages

Unit	Time	CA	Unit	Time	CA
1	133.23	133.23	31	52.6	53.83406397
2	108.3	120.765	32	52.6	53.21703198
3	82.32	101.5425	33	52.5	52.85851599
4	79.3	90.42125	34	51.85	52.354258
5	77.74	84.080625	35	51.67	52.012129
6	72.17	78.1253125	36	51.38	51.6960645
7	72.24	75.18265625	37	51.63	51.66303225
8	72.4	73.79132813	38	51	51.33151612
9	71.24	72.51566406	39	51.6	51.46575806
10	72	72.25783203	40	52.1	51.78287903
11	70.7	71.47891602	41	50.8	51.29143952
12	68.3	69.88945801	42	50.11	50.70071976
13	66	67.944729	43	50.96	50.83035988
14	66.6	67.2723645	44	50.81	50.82017994
15	64.4	65.83618225	45	50.3	50.56008997
16	62.1	63.96809113	46	50.5	50.53004498
17	62.2	63.08404556	47	50.8	50.66502249
18	62.8	62.94202278	48	49.3	49.98251125
19	60.11	61.52601139	49	49.36	49.67125562
20	59.66	60.5930057	50	49.66	49.66562781
21	59.13	59.86150285	51	48.69	49.17781391
22	58.7	59.28075142	52	48.37	48.77390695
23	55.4	57.34037571	53	47.82	48.29695348
24	56.7	57.02018786	54	47.67	47.98347674
25	55.5	56.26009393	55	47.51	47.74673837
26	55.2	55.73004696	56	46.8	47.27336918
27	55.4	55.56502348	57	46	46.63668459
28	56.2	55.88251174	58	42.58	44.6083423
29	55.87	55.87625587	59	42.16	43.38417115
30	54.26	55.06812794	60	41.58	42.48208557

In order to calculate the unit improvement, sample calculations for the Nth (values) unit in table 7 is as follows:

TN = T1 (Nb);

TN = time for the *N*th unit

T1 = hours to produce the first unit

b = (log of the learning rate)/ (log 2) = slope of the learning curve

<u>60% learning rate</u>

- $TN = ($ hours$) (3^b)$

  $= (82.32) (3^{\log .6/\log 2})$

  $= (82.32) (3^{-0.7369656}) = 36.633922$ labor-hours

- $TN = ($ hours$) (3^b)$

  $= (72.17) (6^{\log .6/\log 2})$

  $= (72.17) (6^{-0.7369656}) = 19.27019$ labor-hours

- Etc.

<u>65% learning rate</u>

- $TN = ($ hours$) (4^b)$

  $= (79.3) (4^{\log .65/\log 2})$

  $= (79.3) (4^{-0.6214884}) = 33.50425$ labor-hours

- $TN = ($ hours$) (7^b)$

  $= (72.24) (7^{\log .65/\log 2})$

  $= (72.24) (7^{-0.6214884}) = 21.555551$ labor-hours

- Etc.

<u>70% learning rate</u>

- $TN = ($ hours$) (9^b)$

  $= (71.24) (9^{\log .7/\log 2})$

  $= (71.24) (9^{-0.5145732}) = 22.998331$ labor-hours

- $TN = ($ hours$) (12^b)$

  $= (68.3) (12^{\log .7/\log 2})$

  $= (68.3) (12^{-0.5145732}) = 19.015291$ labor-hours

- Etc.

<u>75% learning rate</u>

- TN = (hours) (2^b)

    = (108.3) $(2^{\log.75/\log 2})$

    = (108.3) $(2^{-0.4150375})$ = 81.225 labor-hours
- TN = (hours) (5^b)

    = (77.74) $(5^{\log.75/\log 2})$

    = (77.74) $(5^{-0.4150375})$ = 39.860778 labor-hours
- Etc.

<u>80% learning rate</u>

- TN = (hours) (2^b)

    = (108.3) $(2^{\log.8/\log 2})$

    = (108.3) $(2^{-0.32193})$ = 86.64 labor-hours

- TN = (hours) (2^b)

    = (82.32) $(3^{\log.8/\log 2})$

    = (82.32) $(2^{-0.32193})$ = 57.7972 labor-hours

- Etc.

<u>85% learning rate</u>

- TN = (hours) (8^b)

    = (72.4) $(8^{\log.85/\log 2})$

    = (72.4) $(2^{-0.2344653})$ = 44.46265 labor-hours
- TN = (hours) (11^b)

    = (70.7) $(11^{\log.85/\log 2})$

    = (70.7) $(11^{-0.2344653})$ = 40.294801 labor-hours
- Etc.

<u>90% learning rate</u>

- TN = (hours) (15^b)

    = (64.4) $(15^{\log.85/\log 2})$

    = (64.4) $(15^{-0.1520031})$ = 42.669382 labor-hours

- TN = (hours) (18^b)

    = (62.8) $(18^{\log.85/\log 2})$

    = (62.8) $(18^{-0.1520031})$ = 40.47197 labor-hours
- Etc.

-

95% learning rate

- TN = (hours) (17^b)

$$= (62.2)\ (17^{\log .85/\log 2})$$
$$= (62.2)\ (17^{-0.0740006}) = 50.435514 \text{ labor-hours}$$

- Etc.

Improvement curves: Table of Units values (60%, 65%, 70%, 75%, 80%, 85%, 90%, and 95% learning rates)

Table 7 Unit improvement factor

Unit	60%	65%	70%	75%	80%	85%	90%	95%
1	133.23	133.23	133.23	133.23	133.23	133.23	133.23	133.23
2	64.98	70.395	75.81	81.225	86.64	92.055	97.47	102.885
3	36.633922	41.58915	46.772606	52.177366	57.797177	63.626349	69.659677	75.892369
4	28.548	33.50425	38.857	44.60625	50.752	57.29425	64.233	71.56825
5	23.742568	28.591886	33.960441	39.860778	46.304847	53.304068	60.869388	69.011322
6	19.27019	23.699804	28.703903	34.307949	40.536659	47.414074	54.963617	63.208147
7	17.217479	21.555551	26.540736	32.21294	38.611523	45.775343	53.742792	62.551826
8	15.6384	19.88285	24.8332	30.54375	37.0688	44.46265	52.7796	62.07395
9	14.108473	18.183324	22.998331	28.620518	35.11773	42.558587	51.01244	60.54934
10	13.193709	17.212507	22.01706	27.688217	34.308711	41.963069	50.73754	60.720021
11	12.076715	15.92964	20.584801	26.13379	32.671258	40.294801	49.104863	59.204641
12	10.942115	14.578811	19.015291	24.35118	30.690357	38.140835	46.814654	56.827777
13	9.9679537	13.404206	17.633502	22.762275	28.902424	36.171201	44.691106	54.589797
14	9.5239542	12.917218	17.128033	22.273482	28.477602	35.871293	44.592233	54.784801
15	8.7528014	11.966271	15.984566	20.929743	26.932027	34.129768	42.669382	52.705317
16	8.04816	11.085238	14.91021	19.648828	25.43616	32.416588	40.74381	50.580838
17	7.7088903	10.692535	14.475527	19.191457	24.98471	32.01053	40.435083	50.435514
18	7.4622019	10.418912	14.191558	18.922324	24.765788	31.88907	40.47197	50.707097
19	6.8635582	9.6430902	13.210962	17.709897	23.295928	30.138625	38.421315	48.341288
20	6.5594723	9.2706082	12.770506	17.207073	22.742863	29.555405	37.83752	47.79762
21	6.2715913	8.913822	12.343244	16.712341	22.189542	28.959656	37.224293	47.20227

22	6.0161516	8.5968235	11.963642	16.273552	21.700739	28.437187	36.69321	46.697974
23	5.4949431	7.8924482	11.035732	15.077927	20.189767	26.560234	34.397187	43.927968
24	5.4502304	7.8667947	11.050028	15.161551	20.382351	26.913581	34.977332	44.817397
25	5.1767751	7.5073993	10.591331	14.591349	19.690503	26.093037	34.025285	43.736559
26	5.0021004	7.2870137	10.323614	14.278155	19.338349	25.714435	33.640214	43.374075
27	4.8825187	7.1438746	10.161747	14.107177	19.174037	25.580245	33.568972	43.409823
28	4.8220381	7.0850745	10.11737	14.096506	19.224519	25.729304	33.865993	43.918326
29	4.671342	6.8915246	9.8779743	13.811113	18.896948	25.368638	33.488034	43.547213
30	4.4247857	6.5533913	9.4274192	13.225712	18.15319	24.442531	32.355848	42.186351
31	4.1870051	6.2247482	8.9860954	12.647792	17.413037	23.513282	31.210028	40.796613
32	4.090176	6.1031287	8.840482	12.482227	17.235968	23.338899	31.059774	40.700877
33	3.9908626	5.9761369	8.6850589	12.300395	17.033622	23.127066	30.856061	40.5311
34	3.8556845	5.7936521	8.4467725	11.998517	16.661829	22.681416	30.336064	39.940954
35	3.7610875	5.6704578	8.2928244	11.813874	16.44976	22.449577	30.097841	39.717008
36	3.6631331	5.5407708	8.1276049	11.611015	16.209761	22.176614	29.801032	39.411849
37	3.607376	5.4737251	8.0528125	11.535584	16.14559	22.14182	29.821577	39.523399
38	3.494009	5.3180576	7.8461376	11.269399	15.812242	21.735309	29.338519	38.964156
39	3.4680858	5.2944584	7.8330435	11.279718	15.865045	21.857492	29.566707	39.346852
40	3.4369612	5.2623055	7.8065767	11.26997	15.888745	21.938671	29.73854	39.653758
41	3.2907698	5.0528603	7.5156822	10.87672	15.369625	21.267769	28.887874	38.593731
42	3.188934	4.9101396	7.3222384	10.622215	15.043706	20.860698	28.391314	38.001699
43	3.1872738	4.9209365	7.3568241	10.697413	15.183435	21.097832	28.769821	38.579074
44	3.1245042	4.8368482	7.2489071	10.56464	15.027115	20.922648	28.585073	38.400134
45	3.0423365	4.7218874	7.0936405	10.361504	14.769046	20.603789	28.201653	37.951531
46	3.005357	4.6763468	7.0417533	10.30824	14.723224	20.579387	28.219352	38.04051
47	2.9756724	4.6416707	7.005627	10.277332	14.708502	20.597517	28.294346	38.205642
48	2.8433477	4.4460571	6.7255109	9.8870962	14.177777	19.890892	27.37115	37.019802
49	2.8038759	4.394788	6.6626284	9.8147761	14.101118	19.819053	27.318706	37.008345
50	2.7792286	4.3663305	6.6337993	9.7919785	14.094853	19.845282	27.40047	37.177651
51	2.6854638	4.2286794	6.4382816	9.5221303	13.731721	19.367515	26.784517	36.39809
52	2.6299087	4.1504956	6.3323776	9.3836187	13.556463	19.152829	26.530062	36.106953
53	2.5637613	4.0550122	6.1993116	9.203869	13.320383	18.85067	26.152566	35.646111
54	2.5207545	3.9956052	6.1207099	9.1040949	13.198936	18.709363	25.996563	35.485179
55	2.4785496	3.9370402	6.0428398	9.0047001	13.077158	18.566517	25.837145	35.318088

56	2.409303	3.8350172	5.8975984	8.8040453	12.807224	18.211952	25.381417	34.743928
57	2.3374293	3.7282242	5.7442289	8.5902128	12.516773	17.826504	24.880518	34.105315
58	2.1360917	3.4139382	5.2697853	7.8943597	11.52149	16.433992	22.969902	31.529055
59	2.0885438	3.344542	5.1721089	7.761231	11.345237	16.206802	22.684312	31.178594
60	2.0344555	3.2642555	5.0570303	7.6012504	11.128782	15.92098	22.315159	30.711446

In order to estimate the learning rate, sample calculations for the rate and average rate in table 8 are as depicted below:

- Units 1 to 2 = 108.3/133.23 = 0.812879982 = 81.29 %

- Units 2 to 4 = 79.3/108.3 = 0.7322253 = 73.22 %

- Units 4 to 8 = 72.4/79.3 = 0.912988651 = 91.3 %

- Units 8 to 16 = 62.1/72.4 = 0.857734807 = 85.77 %

- Units 16 to 32 = 52.6/62.1 = 0.847020934 = 84.7 %

Average = (81.29+73.22+91.3+85.77+84.7)/5 = 83.26 = **83%**

Table 8 Learning rate

Units	Rate	%
Units 1 to 2	0.812879982	81.29
Units 2 to 4	0.7322253	73.22
Units 4 to 8	0.912988651	91.3
Units 8 to 16	0.857734807	85.77
Units 16 to 32	0.847020934	84.7
	Average	**83.26**
	→	**83%**

- Normal time = **avg time *rating factor**

Average (45th unit- 60th unit)

$$\frac{(50.3+50.5+50.8+49.3+49.36+49.66+48.69+48.37+47.82+47.67+47.51+46.8+46+42.58+42.16+41.5}{16}$$

= 47.44375 * 1.2 = **56.9325** as the normal time

Table 9 Unit improvement factor (83%)

Unit	T1	Learning rate	b	nth unit	Time for the nth unit	Unit	T1	Learning rate	b	nth unit	Time for the nth unit
1	133.23	83%	-0.26882	1	133.23	31	52.6	83%	-0.26882	31	20.89694
2	108.3	83%	-0.26882	2	89.889	32	52.6	83%	-0.26882	32	20.71935
3	82.32	83%	-0.26882	3	61.26990017	33	52.5	83%	-0.26882	33	20.50961
4	79.3	83%	-0.26882	4	54.62977	34	51.85	83%	-0.26882	34	20.09378
5	77.74	83%	-0.26882	5	50.43705095	35	51.67	83%	-0.26882	35	19.86859
6	72.17	83%	-0.26882	6	44.58375142	36	51.38	83%	-0.26882	36	19.60803
7	72.24	83%	-0.26882	7	42.81551987	37	51.63	83%	-0.26882	37	19.55885
8	72.4	83%	-0.26882	8	41.3973788	38	51	83%	-0.26882	38	19.18218
9	71.24	83%	-0.26882	9	39.46458431	39	51.6	83%	-0.26882	39	19.2728
10	72	83%	-0.26882	10	38.77177984	40	52.1	83%	-0.26882	40	19.32757
11	70.7	83%	-0.26882	11	37.10868869	41	50.8	83%	-0.26882	41	18.72063
12	68.3	83%	-0.26882	12	35.02020624	42	50.11	83%	-0.26882	42	18.34712
13	66	83%	-0.26882	13	33.12053127	43	50.96	83%	-0.26882	43	18.54068
14	66.6	83%	-0.26882	14	32.76240736	44	50.81	83%	-0.26882	44	18.37222
15	64.4	83%	-0.26882	15	31.09802637	45	50.3	83%	-0.26882	45	18.07827
16	62.1	83%	-0.26882	16	29.47161734	46	50.5	83%	-0.26882	46	18.04323

17	62.2	83%	-0.26882	17	29.04190458	47	50.8	83%	-0.26882	47	18.04579
18	62.8	83%	-0.26882	18	28.87495779	48	49.3	83%	-0.26882	48	17.4141
19	60.11	83%	-0.26882	19	27.23932429	49	49.36	83%	-0.26882	49	17.33892
20	59.66	83%	-0.26882	20	26.66518389	50	49.66	83%	-0.26882	50	17.34983
21	59.13	83%	-0.26882	21	26.0839389	51	48.69	83%	-0.26882	51	16.92062
22	58.7	83%	-0.26882	22	25.57245292	52	48.37	83%	-0.26882	52	16.7219
23	55.4	83%	-0.26882	23	23.84813942	53	47.82	83%	-0.26882	53	16.44733
24	56.7	83%	-0.26882	24	24.1301014	54	47.67	83%	-0.26882	54	16.31356
25	55.5	83%	-0.26882	25	23.36163754	55	47.51	83%	-0.26882	55	16.1788
26	55.2	83%	-0.26882	26	22.99167062	56	46.8	83%	-0.26882	56	15.86002
27	55.4	83%	-0.26882	27	22.84205629	57	46	83%	-0.26882	57	15.51491
28	56.2	83%	-0.26882	28	22.94647528	58	42.58	83%	-0.26882	58	14.29442
29	55.87	83%	-0.26882	29	22.59756176	59	42.16	83%	-0.26882	59	14.08854
30	54.26	83%	-0.26882	30	21.74727478	60	41.58	83%	-0.26882	60	13.83208

8. Results

Figure 1 Learning curve (Cumulative Average & Time)

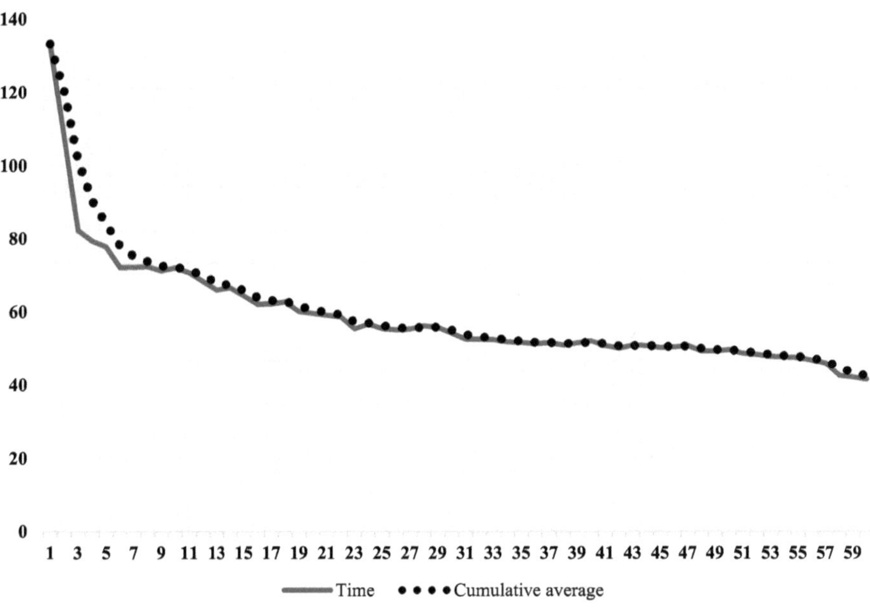

From the above graph, we can deduce that the cumulative averages were getting decreased (133-42); steady line (32th unit - 49th unit).

9. Discussions

From the above data (table 7) it might be argued that the variation in the rate that organizations learn may be due to organizational "forgetting", employee turnover, transfer of knowledge, and the failure to control for other factors, such as economies of scale, when estimating learning curve. Levy (1965) used direct labour training hours to explain different learning rates across workers in a cross-sectional study. Adler and Clark (1991) used longitudinal data on cumulative number of hours spent by workers on training and cumulative number of hours spent on engineering changes [4].

Drawing on more than 200 learning curve studies, Dutton and Thomas (1984) conclude that a learning rate should no longer be treated as a given constant based on past performance, but as a dependent variable influenced by a firm's behaviour [5]. The unit improvement factor in table 9 (time for the nth unit) was improved with the application of the learning rate (83%) and the outcome appears to be enhanced for the 1st unit than for the rest of the units. There are several possible explanations for this observation.

First, airframes firms use the 80 percent learning rate in order to assess the average direct labour hour per aircraft. Then these results are more or less the same to other studies done before in other places in the world where 80 percent learning curve was applied indiscriminately to all stages of airframes production.

Second, based on a review of the literature on learning effects across different organizational contexts, Argote (2013) concluded that the biggest difference in learning rates was between manufacturing and service sectors, with organizations in the manufacturing sector learning at a faster rate than those in the service sector. In addition, the improvement factor curves requires learning how the logarithm model changes each unit; it is initially estimated that 80% gives fairly accurate results. That said, we did not see any evidence of changes in our learning curve; the rate progress ratio falls at 81 to 83% which gives rise to the general assumption of an "80% learning curve" [6].

The level of error in our method would be very low if any because it is based on an exact assignment. The data may be skewed by factors such as the state of mind of the participant at the time of taking the study. For instance, if the participant is under a lot of stress or personal problem, he or she might get difficulty to perform the task under such circumstances. This type of error cannot be eliminated from the process and it is also difficult to measure or estimate its effect on the results.

10. Conclusion

This study has provided an insight into the learning curve; we have answered our questions via a repeatability of our dataset. Besides, the questions that would arise from this conclusion are: do emotional or cognitive factors play a role on the performing task? Why knowledge acquired through learning by doing depreciate?

To determine the factors that affect knowledge depreciation, Bailey (1989) conducted a laboratory experiment in which subjects worked at a repetitive manual task in two separate time periods with a break in between the two periods. In each period the subjects worked for several hours and the breaks ranged from 7 to 114 days. The first period data were used to estimate a commonly used log-linear learning curve model. The model provided estimated completion times of the tasks for the second period. Then the difference between the actual time and the learning-curve-estimated time was considered as the "amount of forgetting" attributed to interruption. Moreover, the difference between the initial task time and the learning-curve-estimated time for the first job in the second period was considered as the "amount of learning" prior to interruption [7].

We conclude that the learning curve can be replicated when it's continuing and permanent; through a logarithm analysis, we can also estimate the learning rate and the improvement curves.

The knowledge and insight gained from such study will add to the literature and be helpful in numerous ways. Therefore, more research on method study is needed in a South African context, not only on engineering discipline fellows. It is plausible that the results will be more or less similar from other countries; this will reveal an important knowledge on technical and the influence of environment on psychological effects toward the study. Moreover, learning is a powerful source of productivity growth; thus a better understanding of learning can enhance any performance.

11. References

[1] L.Argote, 7/8 1996. Organizational learning curve: Persistence, Transfer and turnover, International journal of technology management 11, nos. pp 759-69.

[2] British Standards Institute –BS3375 – Part 2 1993.

[3]T. P. Wright, February 1936. Factors Affecting the Cost of Airplanes, Journal of the Aeronautical Sciences, pp 122-28.

[4] Michael A. Lapré and Luk N. Van Wassenhove. October 2001, Management Science, Informs Vol. 47, No. 10, pp. 1311–1325.

[5] J. M. Dutton and A. Thomas 1984. Treating Progress Functions as a Managerial Opportunity, Academy of Management Review, 9, 235-247.

[6] L.Argote and D. Epple, Feb.23, 1990. Learning Curves in Manufacturing, American Association for the Advancement of Science, Vol. 247, No. 4945, pp. 920-924.

[7] Bailey, C. 1989. Forgetting and the learning curve: a laboratory study. Management Science, 35: 340–352.

Appendices

Appendix A: Ethical considerations

There are some ethical considerations to take note of; we need to ensure that:

- The participation is voluntary
- People cooperate and also remain.

Appendix B: Limitations of Learning curve theory

Following limitations of learning curve must be kept in view:

- All activities of a firm are not subject to learning effect.
- It is correct that learning effect does take place and average time taken is likely to reduce. But in practice it is highly unlikely that there will be a regular consistent rate of decrease. Therefore any cost predictions based on conventional learning curves should be viewed with caution.
- Considerable difficulty arises in obtaining valid data, which will form basis for computation of learning effect.
- Even slight change in circumstances quickly renders the learning curve obsolete; while the regularity of conventional learning curves can be questioned, it would be wrong to ignore learning effect altogether in predicting future costs for decision purposes.

Appendix C: Typical Learning Rates

- Aerospace : 85%
- Shipbuilding : 80-85%
- Complex machine tools for new models : 75-85%
- Repetitive electronics manufacturing : 90-95%
- Repetitive machining or punch-press operations : 90-95%
- repetitive electrical operations : 75-85%
- Repetitive welding operations : 90%
- Raw materials : 93-96%
- Purchased Parts : 85-88%

Appendix D: England Flyer Instructions

Figure 2 Folding Instructions (England Flyer)

Fold along lines 1, 2 and 3

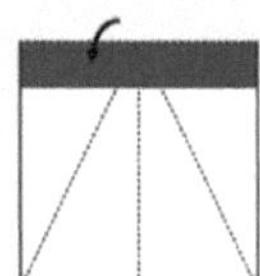

Fold down line 3

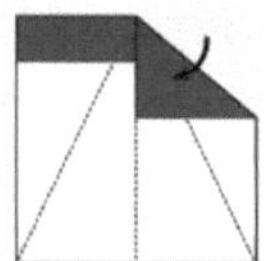

Fold down line 4 on both sides

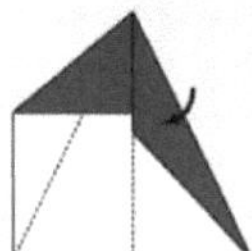

Fold down lines 2 on both sides

The shape should look like this

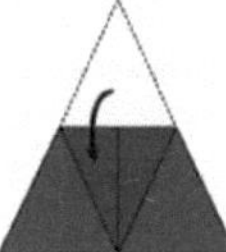

Turn the plane over and fold line 5 and bring the point in

Fold in half along line 1

The plane should look like this

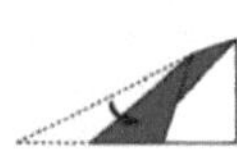

Fold in line 6

Open out the plane. It should look like this

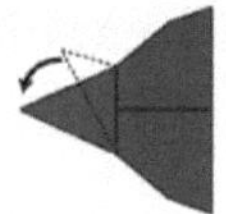

Fold out the nose and fold line 7. Fold in half again on line 1

Fold line 8 into the plane to make a tail

Fold line 9 to make wing stabilisers

Fold line 10 to make the wings and finish the plane

© Copyright The Mighty Eagle

YOUR KNOWLEDGE HAS VALUE

- We will publish your bachelor's and master's thesis, essays and papers

- Your own eBook and book - sold worldwide in all relevant shops

- Earn money with each sale

Upload your text at www.GRIN.com
and publish for free